INVE… …CE

I0893695

Kathy Furgang

# MOUNTAINS

Enslow Publishing
101 W. 23rd Street
Suite 240
New York, NY 10011
USA
enslow.com

# Words to Know

**crust** The outer layer of the earth's surface.

**fault** A break in the earth's crust where large blocks of land shift past each other.

**inspired** To be moved to do something.

**magma** Molten rock underneath the earth's crust.

**mountain** A landform that is higher than the land around it, usually at least 1,000 feet above sea level.

**plate** A piece of the earth's surface that pieces together with others to form the entire crust.

**range** A group of mountains.

**tectonics** The movement of the earth's crust.

**volcano** An opening in the earth's crust where magma pushes through and erupts as lava.

# Contents

# Mountains in the Making

**Mountains** are like a picture-perfect image of nature. We see snowy mountain peaks in photos and on TV. We may see them near our homes. People throughout history have wondered about mountains. They have felt **inspired** to climb them. But what are mountains? How do they form?

A mountain is a landform that reaches at least 1,000 feet (300 meters) above the land around it. The story of a mountain involves movement of the earth's surface as well as the materials inside the earth. It's an amazing journey that takes millions of years.

**Mountain climbers need special gear to climb in high rocky and snowy landscapes.**

Mountains are one of the world's unique land features.

# What's in the Crust?

The earth's outer layer, or crust, is made of seven major plates. Each of these plates has smaller minor plates.

## Plate Tectonics

Mountains form because of the shifting and moving of the earth's thick, rocky surface. The earth's surface is made of large sections called **plates**. Together, these plates make up the outer layer of the planet, called the **crust**. **Tectonics** is the action of these plates as they move, shift, and push together.

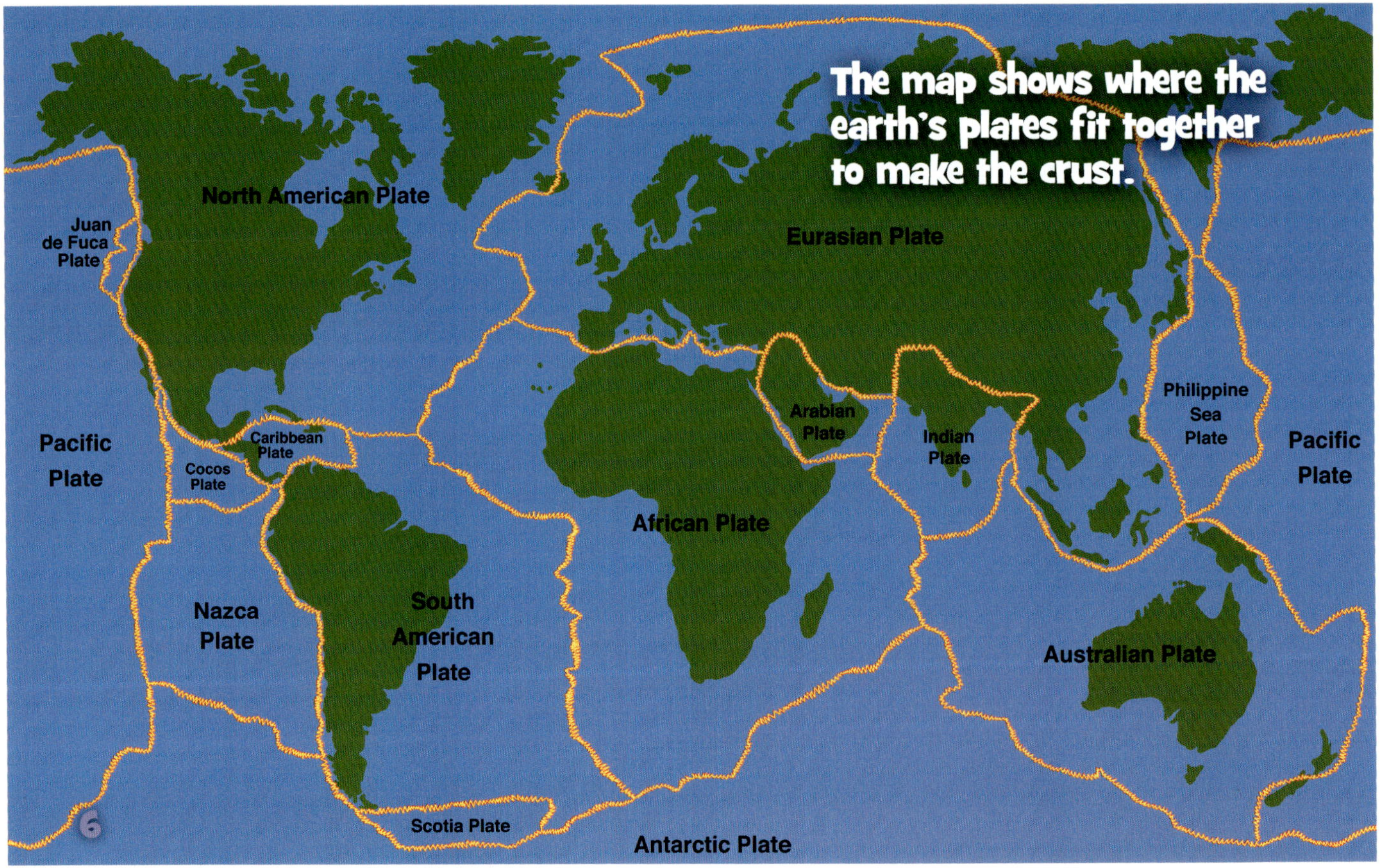


The map shows where the earth's plates fit together to make the crust.

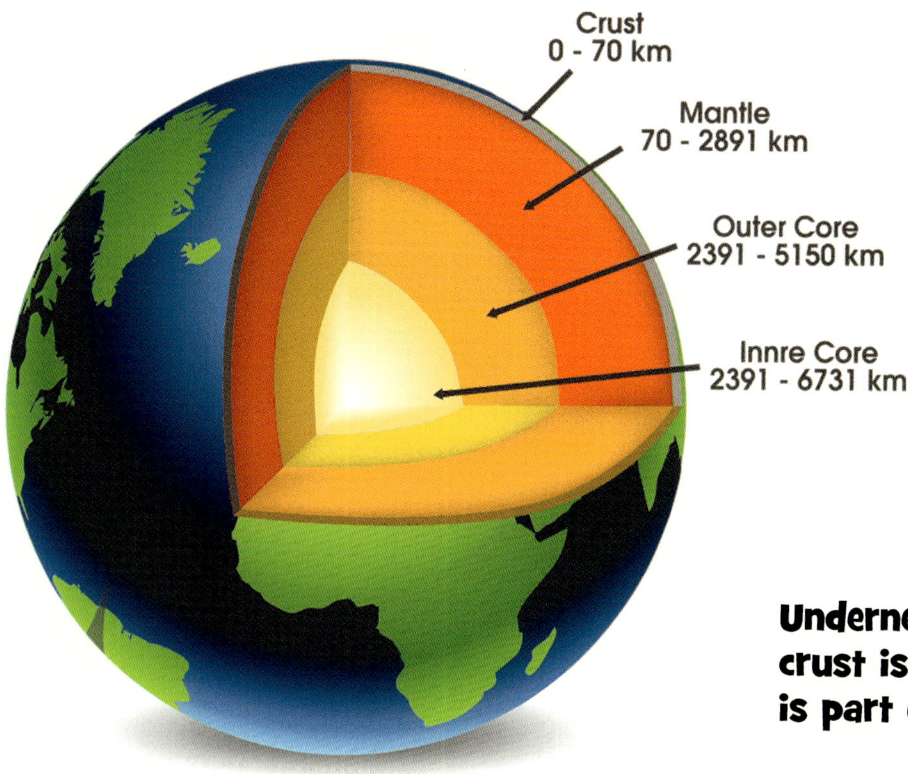


**Underneath the earth's crust is the mantle. Magma is part of the mantle.**

Why do the plates move? Inside the earth's surface, the temperatures are so high that the rock there is completely melted. This material is called **magma**. It heats the earth's crust and makes the plates shift and move. As the plates move, they bump against each other. This is the start of the actions that make mountains.

# Fold Mountains

Mountains form from different types of crust movements. The most common type of mountain is called a fold mountain. To make this kind of mountain, two plates crash together. The land buckles and pushes upward. Think of a sheet of paper that slides into the side of a book. The middle of the paper pushes up. It buckles, or folds, in the middle. This action does not all happen at once. Rocks slide down hills as the crust pushes up and folds. The pushing happens slowly. Over millions of years, the land rises higher and higher. The highest mountain top is on Mount Everest in Asia. It is 29,035 feet (8,850 m) above sea level.

## Mountain Ranges

Fold mountains don't just form one at a time. The earth's plates are long. The entire edge is affected when one plate collides with another. This causes tectonic movement along the whole distance of the plate. This can cause a long mountain **range** with many mountains.

Mount Everest is part of the Himalayan Mountains in Nepal. It is the tallest mountain in the world.

## Mountain Weather

The higher off the ground an area is, the cooler the temperatures will be there. Mountains are so high above the ground that they have different weather. Temperatures are colder than on the ground. It can be snowy and cold at the top of a high mountain but warm and dry at the bottom.

## Story of the Andes

The longest mountain range in the world is made of fold mountains. The Andes are a mountain range in South America. They are more than 4,400 miles (7,000 km) long and 200 miles (300 km) wide. They stretch through more than seven countries.

People have settled in the Andes since ancient times. Today there are modern cities, like Santiago de Chile, near the mountain range.

## Going Downhill

**The land near the Andes is so steep that special farming practices are needed. These methods keep water from rushing downhill and away from crops.**

The huge range began forming millions of years ago when two plates collided. One plate pushed down below the other. Over time, earthquakes and **volcanoes** continued to shape the land. Even today, the area has many active volcanoes. There are more active volcanoes in this area than anywhere in the world.

# Other Mountain Types

While fold mountains are the most common, mountains form in other ways, too. All of them form based on tectonic activity, close to where plates meet.

## Fault-Block

A **fault**-block mountain forms when one plate is broken into chunks, or blocks. These blocks are forced up or down when they are pushed together. Over time, the shifting forms higher and steeper land.

## Dome

Another mountain type is called a dome mountain. When magma under the crust pushes up, the crust bends upward. It forms a dome shape. The magma does not

The Sierra Nevada in the United States are fault-block mountains. Half Dome (on the left) is part of the Sierra Nevada range. It is a dome mountain.

push through the surface, however. Instead, it cools and hardens into rock.

## Volcanic Mountains

Like dome mountains, volcanic mountains form because of magma. With volcanic mountains, however, magma pushes through the earth's crust. It travels through a passage, called a vent. Magma that has reached the surface is called lava. Volcanoes erupt with lava, smoke, and ash. The lava then cools into rock.

Mayon Volcano in the Philippines, before and during an eruption. Magma pushes above the surface through a vent down the center of the mountain. It erupts as lava, smoke, and ash.

## What Kind Is It?

**Scientists place volcanoes into three groups based on how often they erupt. Active volcanoes erupt regularly. Dormant volcanoes erupt only once in a while. Extinct volcanoes no longer erupt. Magma can no longer escape from them.**

Some volcanoes can go hundreds or thousands of years between eruptions. Others erupt often. Volcanic eruptions change mountains. Lava cools into rock after an eruption. This builds the mountain up. Other times, a powerful eruption blows apart some of the mountain. This breaks the mountain down.

# Mountains Are Always Changing

Once a mountain is formed, it does not stop changing. Weather, such as wind, rain, ice, and changes in temperature, breaks apart rocky mountain surfaces. Rushing water moves rocks in mountain streams. Melting ice and waterfalls also change mountains over time.

Earthquakes and natural plate tectonics also continue to shape the earth's surface.

Mountains even affect the weather. Strong winds and ocean storms may strike one side of a mountain near a coastline. The other side of the mountain is more protected. On the west coast of North America, mountain ranges get more rain and snow on the west side than on the east side.

The Olympic Mountains are located in western Washington State.

Olympus Mons is a volcano on Mars. It is about the size of Arizona and reaches 16 miles (25 kilometers) high! It is 100 times larger than Mauna Loa, the earth's largest volcano.

## Wild Wind

**The west side of North America's Pacific mountain system is called the windward side because it gets more wind than the eastern side.**

## Mountains Underwater and on Other Planets

Tectonic plates cover the entire planet. That means they are also on the ocean floor. In fact, some of the most dramatic mountain ranges are under the ocean. There are even more volcanoes underwater than on land.

Scientists have found that the earth is not the only planet with mountains. In fact, the tallest mountain in our solar system is on Mars. Some volcanoes there are up to 100 times larger than volcanoes on the earth.

There's always something new to learn about mountains. Why? Because from the bottom of the oceans to the top of the tallest mountain, the earth is always changing.

# Activity: Candy Mountains

You can use a chocolate bar to model how a mountain forms. You will need:

- mini chocolate bars with a soft caramel center
- a butter knife

1. Carefully cut a thin slit across the top of a chocolate bar with a soft caramel center. Be sure to only cut through the top layer. The chocolate top forms two "plates" in the earth's crust. The caramel is magma inside Earth's crust.
2. Show how plate tectonics can form a mountain. What can you do to the chocolate bar to model the movement of plates? What happens to the magma? What kind of mountain did you form?
3. Try the activity again. How can you change the candy bar to form a dome mountain? Hint: Do not cut the chocolate bar.

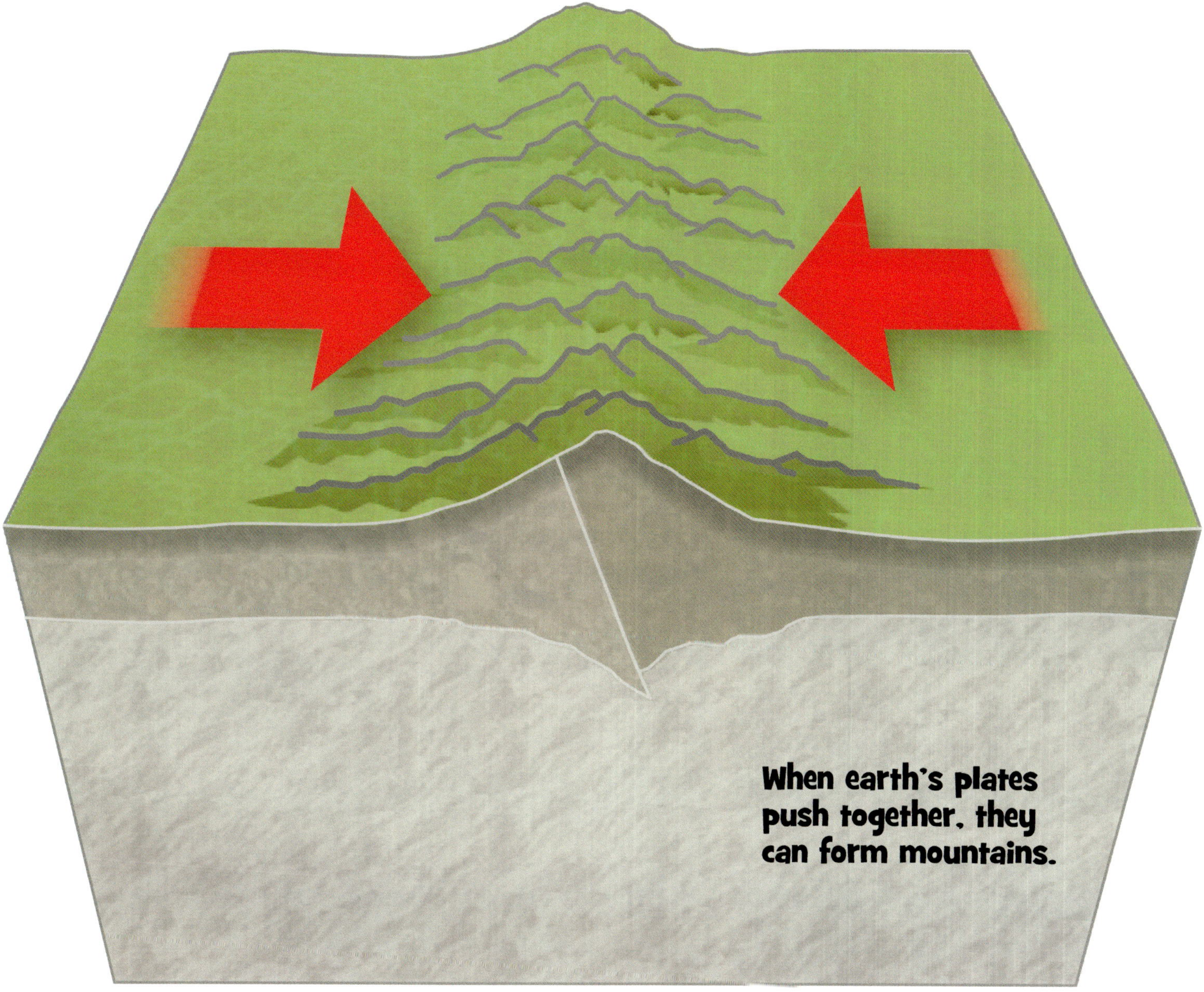

When earth's plates push together, they can form mountains.

# Learn More

## Books

Best, B.J. ***How Are Mountains Formed?*** New York, NY: Cavendish Square, 2017.

Francis, Sangma. ***Everest***. London, UK: Flying Eye Books, 2018.

London, Martha. ***Mountains***. Mendota Heights, MN: Focus Readers, 2018.

## Websites

**Kids Discover: Himalayas**
*www.kidsdiscover.com/spotlight/himalayas-kids/?mc_cid=cb16463267*
Learn more about the Himalayas, the mountain range that includes Mount Everest.

**Mountains: National Geographic Kids**
*kids.nationalgeographic.com/explore/nature/habitats/mountain/#Denali-mountain.jpg*
Discover how mountain climates change the higher you climb. Learn about different animals that live in mountain habitats.

# Index



Published in 2020 by Enslow Publishing, LLC.
101 W. 23rd Street, Suite 240, New York, NY 10011

Copyright © 2020 by Enslow Publishing, LLC.

All rights reserved.

No part of this book may be reproduced by any means without the written permission of the publisher.

**Library of Congress Cataloging-in-Publication Data**
Names: Furgang, Kathy, author.
Title: Mountains / Kathy Furgang.
Description: New York, NY : Enslow Publishing, 2020. | Series: Investigate earth science | Includes bibliographical references and index. | Audience: Grade level K-4.
Identifiers: LCCN 2018049716| ISBN 9781978507456 (library bound) | ISBN 9781978508712 (pbk.) | ISBN 9781978508729 (6 pack)
Subjects: LCSH: Mountains—Juvenile literature.
Classification: LCC GB512 .F87 2020 | DDC 551.43/2—dc23
LC record available at https://lccn.loc.gov/2018049716

Printed in the United States of America

**To Our Readers:** We have done our best to make sure all website addresses in this book were active and appropriate when we went to press. However, the author and the publisher have no control over and assume no liability for the material available on those websites or on any websites they may link to. Any comments or suggestions can be sent by e-mail to customerservice@enslow.com.

**Photos Credits:** Cover, p. 1 FROGtograph/Shutterstock.com; pp. 3, 5 Andrew Mayovskyy/Shutterstock.com; pp. 3, 8–9 Daniel Prudek/Shutterstock.com; pp. 3, 11 Jose Luis Stephens/Shutterstock.com; pp. 3, 14–15 Gary C. Tognoni/Shutterstock.com; pp. 3, 16 (bottom) Glenda Esperida/Shutterstock.com; p. 4 yanik88/Shutterstock.com; p. 6 Peter Hermes Furian/Shutterstock.com; p. 7 Webspark/Shutterstock.com; p. 16 (top) suronin/Shutterstock.com; pp. 18–19 Jakub Skyta Photography/Shutterstock.com; p. 20 Ron Miller/Stocktrek Images/Getty Images; p. 23 Dorling Kindersley/Getty Images; cover graphics blackpencil/Shutterstock.com.